50 Things to DO with a BOOK

Also by Bruce McCall

All Meat Looks like South America

The Last Dream-o-Rama

Thin Ice

Sit! The Dog Portraits of Thierry Poncelet
(with Thierry Poncelet)

Zany Afternoons

50 Things to DO with a BOOK

(Now That Reading Is Dead)

BRUCE McCALL

HarperCollins books may be purchased for educational, business, or sales promotional use. For information, please write: Special Markets Department, HarperCollins Publishers, 10 East 53rd Street, New York, NY 10022.

FIRST EDITION

Designed by Nicola Ferguson

Library of Congress Cataloging-in-Publication Data
McCall, Bruce.
50 things to do with a book : (now that reading is dead) / Bruce McCall.—1st ed.
p. cm.
ISBN 978-0-06-170366-9
1. Books—Humor. 2. Books and reading—Humor. I. Title: Fifty things to do with a book.
PN6231.B62M38 2009
818'.5402—dc22
2008053683

09 10 11 12 13 OV/RRD 10 9 8 7 6 5 4 3 2 1

For Johannes Gutenberg

INTRODUCTION

Hotel rooms now provide Gideon Bibles only as pay-per-view TV channels. Librarians recently thrown out of work are forced to take jobs assembling Kindles in Amazon.com basement facilities where books were formerly stored. The Cassandra Report, the bible of the book-publishing business (available only on the World Wide Web), forecasts that more than fifty warehouses across the U.S., long used as book storage and shipping centers, will shortly be converted to video-game facilities. The Barnes & Noble retail book chain is widely rumored to be studying conversion of many of its larger stores into more profitable "Noble Barns"—mini-storage units. Rare-book collectors are switching to classic Betamax movie videos of the 1970s and 1980s.

Meanwhile, any publishing executive or editor still

employed today will admit that being paid in books instead of currency, though reducing their firm's unsold inventory and representing once-in-a-lifetime windfalls for bibliophile employees, can create morale issues that not even free copies of *How to Get Rich without Brains!* can easily resolve.

Dire omens indeed, in line with a recent survey that found that more than half of all Americans didn't read a single book in the previous year—doubtless a conservative figure, because everybody lies about their reading habits. The trend toward a bookless society is gaining almost daily as a TV-besotted, iPhone-bedazzled, time-starved, speed-crazed populace becomes ever less willing to seek information and entertainment by concentrating their minds on endless lines of type on even more endless page after page while sitting in a chair or lying in bed, wearing earphones.

In brief: the end of books, as man has known and loved them since Gutenberg's movable-type breakthrough of 1439, is nigh.

That's why *50 Things to Do with a Book* may well be not only the last, but also the most timely, book ever published.

Its inspiring message is that the end of the book as an information-input tool portends *opportunity* as well as intellectual teeth-gnashing. Instead of just sitting there reading books, which is rapidly becoming as passé as a Jane Fonda workout video, there are literally thousands of ways to amortize all the time and money you've spent on all those books on your shelves, coffee table, and nightstand, or in the corner under the bed at the summer cottage—and to help yourself forget the death of the six-century-old book tradition—by doing other things with them. Fun, exciting, adventurous, creative things.

Seen in this light, that familiar hefty rectangular object with a flat, smooth surface on two sides is suddenly seen to offer enough dazzling new interactive possibilities to—yes, fill a book. This book.

And the upside keeps going up. Let's face facts: it may even be a *relief* to toss that Theodore Dreiser novel you never quite got around to reading, or that history of Dutch agriculture you started reading but couldn't finish, without shame. You're not a moron: you're a with-it American on the cutting edge of a worldwide trend. Think, too, of the vast new space suddenly opened up on living room shelves, tabletops, and throughout home and office for your collection of Franklin Mint car models or antique can-openers. Savor the idea that that insufferable book bore holding forth at cocktail parties on the hidden meanings in Danielle Steele will be silenced forever. No longer will your kids have to lug book bags as heavy as boulders to and fro from school.

And ad infinitum, bookless-benefits-wise. Feel better? Imagine how you'll feel after you've finished discovering *50 Things to Do with a Book!*

50 Things to DO with a BOOK

Books on ancient Egypt are the starting point for a fascinating project. Use them—you'll need up to a million volumes—to build your own Pyramid of Cheops. For maximum authenticity, build the pyramid close to a stream or river and have the books loaded into a rowboat or punt way upriver. Pay slaves to sail the cargo down, unload it, and lug the books to your site. (For absolute maximum authenticity, be sure to whip your slaves.) Sunny days are always best when in ancient Egypt. Be sure to study the pyramid on the back of a U.S. dollar bill to make sure yours is the correct four-sided triangle shape.

Polar exploration books can make for a lively outdoor activity. Gather up a hefty selection, tear out all the pages, and rip them into shreds the size of, say, cornflakes. Place the thousands of shreds in a pile in front of a powerful fan, then walk a hundred paces or so away. Have a pal turn on the fan to create a fierce blizzard and start trudging toward it. You'll feel like Scott of the Antarctic!

Membership in a reputable skeet club can be expensive, and even more so if you have to rent a gun. Why not start your own backyard skeet club, exchanging your piles of useless books for the regulation clay pigeons? Get Junior down from his room and have him hide in a bush with a few dozen volumes of poetry or a set of old law books; meanwhile you take up position with your .22 rifle. Ready?

Shout "Book!" every five seconds and start firing away as Junior lobs one book after another in a high arc from his hiding place. If you keep missing your flying targets, tell Junior to go back up to his room until he learns to throw like a little man and not a sissy girl.

Fig. 1
Fig. 2

Cut a circular baseball-sized hole in the center of a thick book, all the way through to the back cover, which must be left intact. Then glue two, three, or four catgut strings eight to ten inches long so that they run across the hole. Buddy, you've got a *bookjo*! So start strummin'!

Those redundant books can make a fascinating bedtime sport. You and your spouse get into bed. While she snuggles down to go to sleep, you gather up half a dozen books and start juggling them without moving from a sitting position. You should get the hang of it in a couple of hours . . . but! Every time you curse out loud and wake the wife is one demerit point. Every time an errant book knocks something off your nightstand is two demerit points. Every time a book glances off the Little Lady or conks her on the head is three demerit points. You can't switch off the light and go to bed until you've erased all demerit points.

Why not build yourself a pair of *elevator books*? It's simpler than it sounds. Get a roll of duct tape and bind the flat sides of two books of exactly equal thickness—twin copies of *Paradise Lost*, for example—to the underside of your feet, shod or shoeless. You'll gain up to five and a half inches of height. Better yet, the distinctive shuffle required to move with two hefty volumes underfoot will make you the focus of all eyes at the cookout, the college reunion, or the convention!

Lassie

Hand your pooch a—yes—dog-eared copy of *Lassie* or *Rin Tin Tin* or *Bob, Son of Battle*—any popular canine novel. If your dog refuses to eat the book, spread lard or chicken fat all over it. He'll gobble it right down!

A Finnish-language book can be exchanged for a genuine pair of reindeer antlers in Macao, the Hindu Kush, and, of course, Finland—but American bookstores today are hypersensitive to returns of all foreign-lingo volumes and will probably give you the bum's rush. No problem: try the Finnish consulate in your community. Most Finnish consular offices provide a return bin, usually within fifty feet of the entrance, that allows you electronically to punch in the price and push a button to receive your antlers. (Sorry, Finnish currency only!)

SUOMI
KAKKAKININI
VELIP
IMI POKKO NIXAS
UPA SUOMA
SIKI TIKI
TAKI
YEK
NEK

Using chintzy lingerie, tightly wrap a dirty old book—"dirty" as in dirty old man—and then drive your car fast past the home of anyone who's ever disappointed or double-crossed you in love and hurl it through an open window. It's no longer a book: it's a tart editorial comment and a stinging rebuff, tinged with sweet revenge.

327

A rollicking Jack London seafaring book can be your ticket to aquatic adventure. Take it outdoors during a heavy rain—and be sure also to bring an antique Victorian-era "bookbrella" as favored by the late Queen Victoria for her outdoor reading. The very crepe-paper construction that makes the bookbrella so delightfully light and portable turns out to be a meteorological two-edged sword, quickly disintegrating under even a light shower into a soggy mush and turning the book being held beneath it into mush, too. Perhaps this explains why Victoria seldom ventured outdoors!

3 Goons
MOVING

Ernest Hemingway fans like to throw their copies of Papa's works into ravines or deep forests and go after them with resolute calm, just as their hero went hunting for Cape buffalo or baby antelopes. Imposing a strict time limit for the retrieval adds to the tension, and thus the fun. For even more fun, seed the area with a live lion.

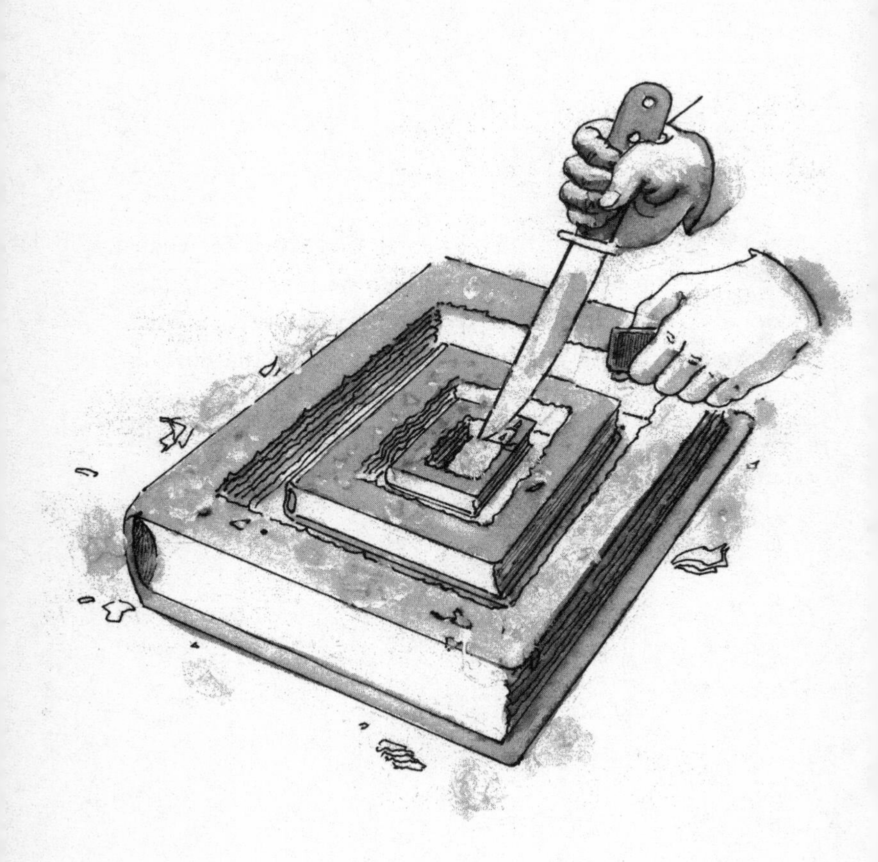

Carve out the insides of a big book and put a smaller book in the space you've created. This "Russian doll" concept can be repeated indefinitely until the last book is as tiny as this.

Here's a rousing book use for the young 'uns: book bombing! A book is dropped from above to hit a circled area on the pavement below, while the bombardier moves overhead on a unicycle or—better yet, because of the greater height and more authentic bombing sensation—stilts.

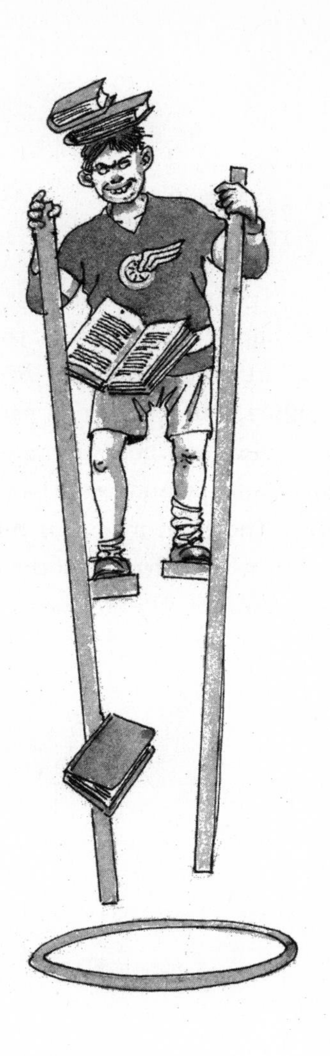

Making sure that you do *not* read the title beforehand, bind any book with packing tape in a crisscross pattern, toss it in a swimming pool or canal, then dive in and try to remove the tape and read the title before you run out of oxygen. The great magician Houdini enjoyed this as an after-dinner hobby, even though historians tell us that Houdini was barely literate in English.

Doorstops are so *obvious*: use your creative imagination to find other practical household uses for all those now surplus books. Smallish paperbacks make ideal beverage coasters, for example, or serving trays for delicate watercress sandwiches. A good solid book is an ideal platform—level, smooth, and sturdy—for hobbies such as flea circuses and, if you have books in French, for as many matchstick Eiffel Towers as you have *livres en française*. Ten stacked copies of the *Encyclopaedia Britannica*, meanwhile, make one swell lampstand, while thirty stacked copies serve as a surefire conversation piece: "Hey, what the heck are you doin' with them thirty *Encyclopaedia Britannicas*?"

MY FAVORITES

Want to test the literary IQ of your friends and family? Write "My Very Favorite Book" on the flyleaf, put it in a prominent place on your bookshelf, and then see who's first to try to steal it. If nobody tries, choose a new favorite book, until you're down to paperbacks with die-cut or embossed foil covers and an oversized title. If nobody bites, turn your energies to finding new friends. Librarians make a good bet.

Blenders and books seldom mix, but if your blender has a supersized bowl and heavy-duty blades—well, the fun's already begun! Note: when you've finished—which shouldn't be more than a few seconds after you start—the ground-up pages make a perfect carpet for that hamster cage. A big enough book makes paper carpet enough to fill the cage of that bull white rhino that Junior just brought home.

Those now hopelessly redundant Book-of-the-Month Club deliveries can be reversed on the spot: hand the packages right back as generous impulse gifts to your favorite postal carrier or UPS man when he arrives at your place with another load. Now the unread-book glut becomes his problem!

Next time you find yourself in a remote jungle in India, hunt down a mature cobra. (Careful, he can bite!) Bring him back to the U.S. and train him to balance several books on his flat head. Some folks like to play "If They Asked Me, I Could Write a Book" or some other biblio-tune on a flute, and others make sure that all the books are by Rudyard Kipling, the immortal bard of the British Raj in India, but those features are strictly optional: the main thing is to draw a crowd of curiosity seekers and charge them as much as the traffic will bear for a five-minute viewing.

BIBLIOPHILE
COBRA
$1 PER VIEW

Using a big bit, drill a hole through the center of a James Gould Cozzens novel and see if you can fill in from memory all the words you've obliterated. (Not kosher to use a Cozzens novel you've read before!)

Dress up that empty wall: find fifty identical copies of the same thinnish hardcover book—children's books by celebrities are ideal—and glue them to the wall end-to-end in a horizontal line to form a "book tile." Paint to taste. Not for the shower!

Pull the spine of a fat book apart so that you have two or three hundred loose pages. Throw the pages in the air, and without looking at the page numbers, reassemble the book. Proper original order is a no-no. By mixing up the pages, you could create a whole *new* literary work and qualify for a prize, a grant, or a summer at a writers' retreat.

Books by the beloved poet Robbie Burns make a perfect excuse for a good old-fashioned Scots-style sporting game of "tossing the caber"—heaving an armful of Robbie's books instead of a giant pole as far as you can. Recruit five or six pals and compete to see who can toss the farthest. The one with the shortest throw has to eat a bowl of haggis!

Don't just haul all your old James Bond books to the local recycling center: don fancy dress, head for any fine amusement park, and jump, leap, dash along atop the bumper cars, swing on poles, climb around the outside structure of the Ferris wheel, and generally cavort athletically while your armload of 007 paperbacks fly like so many defused nuclear mini-bombs snatched from the villain's lair with milliseconds to spare.

Then when the security cops come to take you away, murmur "Bond. James Bond."

After dunking a bunch of moldy old novels in the bathtub for fifteen minutes and letting them dry so the pages stick together, give one book to each person in your gang and have them try pulling the pages apart without tearing them. (Best in summer cottages on rainy weekends.) Only *afterward* should you announce that anybody who's ruined one of your treasured old literary masterpieces must pay.

Rent a powerful chain saw and chop your books into basic geometric shapes: circles, triangles, squares, hectagons, pentagons, diamonds, rhomboids, and the like. These are such fundamental elements of design and even nature that at first it may seem like all you've done is create an unholy splatter of paper fragments—but you'll surely find some useful purpose for them. Hint: ever thought about delivering a *geometry lecture?*

Raid your local library for unwanted or damaged books and start piling them up in any open, flat space to build your own personal stairway to paradise. You'll likely need several thousand books.

Mom, Dad, Sis, and Junior each builds his or her own individual "Bookhenge" in the backyard using nothing but books, piling them into circles of columns that resemble the actual Stonehenge as much as possible—without using any visual reference material. Then invite a friend who's a professor of Druid history over to judge whose Bookhenge is the winner.

Fortunate enough to own a sprawling ranch? Get your ranch hands to collect huge piles of books, pile them in the buckboard, and then head off to a wide-open stretch of prairie. Important: shoo away all grazing livestock in the vicinity. You've already guessed what comes next: a miniature Great Wall of China constructed of books laid end to end and stretching across the landscape in either direction as far as the eye can see. Depending, of course, on how many ranch hands you have and on their collective sense of humor! Make sure you leave at least one of the boys behind to repair the inevitable breaches and collapses as Old Man Winter wreaks his havoc.

Drill two oil wells in an open space. Place piles of books over the wellheads, then take off in a hot-air balloon and maneuver the balloon between the two derricks. At your signal, have a friend dynamite the wells so that twin oil gushers belch high into the air—taking the books with them! You've already placed bets on which gusher will blow the books highest, and from your vantage point in the balloon you can easily judge the winner.

If the gusher you bet on doesn't win, lie to your friend below and say it did. You didn't go to all this trouble to lose money!

BOOKS

Books are highly water-absorbent—particularly pulp fiction. If you live in a Midwest area near a major river or other body of water frequently plagued by seasonal floods, be a community hero and turn those superfluous volumes in your home library into substitutes for sandbags to shore up the banks or the levee as a last defense against the next catastrophic inundation. "Book Worm Turns to Man/Woman of Action!"—the next post-flood local newspaper headline could be about you!

Exactly duplicating the angle of burial, insert up to fifty books in the ground in your back (or front) yard, modeled after Texas's famed Cadillac Ranch, and see if anybody comes—especially if you charge admission.

Cookbooks seldom make a satisfying meal, but to identify them among all the look-alikes on the kitchen shelf, why not wrap your favorite vegetable books in lettuce, wrap your favorite Easter-recipe books in sliced ham, and so on? The instant ID will be very handy!

Big Somerset Maugham fan? Here's a way to memorialize *The Razor's Edge*: glue an actual razor blade to the page of a book so that just the sharp sliver of the edge sticks out. The unwary finger trailing over the pages should get a real feel for the title and its poetic intent!

RAZOR'S
EDGE
ET MAUGHAM

If you're a bookish person of deep faith, lay your library of precious religious tomes in two long parallel lines about three feet apart—then gather family and friends and make a run for it through the channel thus created. You and the gang will feel what Moses & Co. felt as the Red Sea parted in their flight from Egypt. For best results, stage outdoors.

TO KILL A MOCKINGBIRD
TO KILL A MOCKINGBIRD
TO KILL A MOCKINGBIRD

"Kill a Mockingbird" by hiding in the bushes and flinging a copy of this famous book at the first mouthy little avian to appear. A stuffed bird does just as well. A badminton bird is a last resort.

History buff? Stage a "Decline and Fall of the Roman Empire" in your own bedroom: pile the whole of Gibbon's monumental work atop the bookcase, then gently shove one volume against the next. The resulting domino effect should trigger a true decline and fall, with the books cascading into a heap on the floor.

Calling all military hobbyists! The front cover of any large-format art or interior-decoration book, laid flat and spread with a thick layer of peanut butter to resemble battlefield mud and dotted with realistic-looking 1/42nd-scale artillery and tanks and with scattered dead soldier figures laid on their sides (or for even greater realism, chopped in half), can be a stirring diorama dramatizing your hatred of war and its terrible costs. Be patient: the peanut butter will turn rancid, of course—but only for a few days, tops, before it hardens into a perfectly odorless permanent landscape of hideous carnage.

Build a miniature backyard Thornfield Manor out of your old *Jane Eyre* and other Charlotte Brontë novels, pour gasoline on one corner of the "building" when it gets dark, and then light it with a match. See who comes running as the flames rise higher; you could make new friends and later take turns immolating *The Fall of the House of Usher* and other classics featuring the destruction of buildings by fire.

Who's better, Updike or Roth? Finally settle the matter before it doesn't matter anymore with . . . Fightin' Writer Kites! Tie an Updike novel to the tail of one kite while a friend ties a Roth masterpiece to another—or vice versa, depending on tastes—then let 'em soar and start battling. The first kite to have its tether severed will nosedive earthward, and the dangling book attached will crash with a thud, the loser! Updike versus Roth is just one example: Sartre versus Camus, Mary McCarthy versus Lillian Hellman—the list of competitors can be as long as your personal literary hit list!

PHILIP
ROTH

Build a tabletop model of the Himalayas with your own supply of books and those of neighbors, too, if needed. Don't be afraid of a jumble of slabs and sharp edges—Himalayan mountaineers aren't! The *Rand-McNally World Atlas* makes a great Everest. Then sprinkle confectioners' sugar over the "peaks" to create the effect of snow and, for a final meteorological touch, aim a high-powered fan at your mini–mountain range to duplicate those fierce Himalayan winds. Some hobbyists like to catch and sedate a mouse to simulate oxygen deprivation and have the mouse try climbing to the top like a rich American dilettante. Don't worry if it expires on the way: that's Everest!

As with deconsecrating a church, decommissioning books deserves a formal act or ceremony for the serious former book fan. Find a nice empty patch of grass on the lawn of an abandoned library and conduct a funeral. Have all attendees dress as their favorite author, for both verisimilitude and just plain fun. If those authors were funny, like S. J. Perelman or Saki, the funeral doesn't have to be morbid at all! Reading a eulogy would be illogical in view of what you're burying, so get a portable DVD player and run a graveside service plucked from a movie, with the sound turned off while you vamp your eulogy, karaoke-style. There are lots of such movie burial ceremonies to choose from.

Here's an educational exercise: grab one or two books on Arctic exploration and stick them in the freezer compartment of your fridge, leaving them there to frost up for a week or so. Remove the books and immediately press them hard against the cheeks of friends who drop by. Keep pressing until their cheeks are numb. Your friends won't soon forget their firsthand feel of the agony of frostbite as suffered by those heroic Arctic explorers.

Get a biography of Leni Riefenstahl and another of Adolf Hitler and rub them briskly together. Do they burst into flame? If so, you may have just proved a long-standing historian's theory!

Submerge yourself in calm, clear water with a copy of Jules Verne's *Twenty Thousand Leagues under the Sea*. Does it read more vividly underwater? Can you better identify with the story line? After you resurface, decide if specially waterproofed copies would sell, and if so, contact the Verne estate, suggesting a fifty-fifty copublishing deal.

The notorious early-twentieth-century British *Yellow Book* makes a spectacular and natural Yellow Brick Road for the *Wizard of Oz* fan. But considering that the *Yellow Book* came out a hundred or so years back and was relatively scarce even then, you may have some trouble rounding up the two hundred to four hundred copies needed to pave the way. Be sure to seek out flat ground for your road to ease construction; Kansas, Iowa, and sections of Nebraska are ideal.

Send any recent book on African geography over Victoria Falls and have a quick-witted surveyor friend establish (a) how far it falls, (b) how fast it reaches the bottom, and (c) the condition it's in. Then have a certified expert calculate the replacement cost, because it will have already vanished downriver in a swirling whirl of foam.

Build a sturdy catapult outdoors and load its capacious ammunition cradle with, instead of rocks, a collection of dim-bulb celebrity bios, ghostwritten political memoirs, and stupid cookbooks—all the crappy books you can find that greedy publishers have foisted on the buying public over the decades. Calculate the exact location and distance of the nearest strip mall and aim your catapult load accordingly. Cut the restraining ropes and run.

Make a mattress by duct-taping a bunch of old coats and other sturdy outerwear into a solid, roughly rectangular shape. Cut a hole in one end and stuff in fistfuls of shredded paper from now-useless books until the mattress is a fat, fabric-clad moonscape of lumpy bulges. Important: before you shred the books, make sure you've jotted down their titles, then add up to a hundred *bogus* titles. Next morning, hand your overnight guests who just used the mattress the complete list and have them check off the titles of the book contents that they *feel* they slept on. Every correct answer earns five points—but a score of less than a thousand points means your literary-dunce guests have to miss breakfast and make all the beds in the house.

Paint the front cover of a Kindle-sized book to look exactly like a Kindle, pasting a piece of celluloid in place to look like a screen and using the buttons off the sleeve of any good men's suit jacket to imitate the controls. Attend a literary event of great magnitude—say, the Nobel Prize awards for literature—and place your fake Kindle on a chair seat before the proceedings begin. When the occupant discovers the "Kindle" and picks it up, point him out to the crowd and jump up and down while bellowing a hysterical denunciation of this traitor to the immortal tradition of the clothbound printed book.

ABOUT THE AUTHOR

Bruce McCall's written and illustrated satire is a familiar part of the *New Yorker, Vanity Fair,* and other major publications, and he has published two books of collected humor, *Zany Afternoons* and *All Meat Looks like South America*, as well as a memoir of growing up Canadian, *Thin Ice*. He has also just published his first children's book, *Marveltown*. Bruce McCall lives in New York City.